AF719684

THE FUTURE OF AI

AI and the Military

Donna McKinney

ReferencePoint Press

San Diego, CA

© 2027 ReferencePoint Press, Inc.
Printed in the United States

For more information, contact:
ReferencePoint Press, Inc.
PO Box 27779
San Diego, CA 92198
www.ReferencePointPress.com

ALL RIGHTS RESERVED.
No part of this work covered by the copyright hereon may be reproduced or used in any form or by any means—graphic, electronic, or mechanical, including photocopying, recording, taping, web distribution, or information storage retrieval systems—without the written permission of the publisher. No part of this book may be used or reproduced in any manner for the purpose of training artificial intelligence technologies or systems.

LIBRARY OF CONGRESS CATALOGING-IN-PUBLICATION DATA

Name: McKinney, Donna B. (Donna Bowen) author
Title: AI and the Military / by Donna B. McKinney.
Description: San Diego, CA : ReferencePoint Press, Inc, 2027. | Series: The Future of AI | Includes bibliographical references and index.
Identifiers: LCCN 2025048178 (print) | LCCN 2025048179 (ebook) | ISBN 9781678212964 library binding | ISBN 9781678212971 ebook
Subjects: LCSH: Artificial intelligence--Military applications--United States
Classification: LCC UG479 .M44 2027 (print) | LCC UG479 (ebook) | DDC 355.40285/63--dc23/eng/20260206
LC record available at https://lccn.loc.gov/2025048178
LC ebook record available at https://lccn.loc.gov/2025048179

CONTENTS

AI Finds the Target

In January 2024 militia fighters with the Islamic Resistance in Iraq sent an aerial drone to attack a US military camp in Jordan. Three US soldiers died and forty-seven were injured in the drone attack. The US military responded to that attack by bombing Iraq and Syria. The United States used a type of artificial intelligence (AI) called computer vision to identify the targets it bombed. Researchers can create computer vision algorithms, those instructions that the AI technology uses to perform various tasks, so computers can "see" and understand features or patterns in images, allowing them to identify a person, place, or object. Face ID on smartphones is an example of computer vision AI at work in everyday life. The identification algorithms that helped the military find its targets in Iraq and Syria were developed for a US Department of Defense (changed to US Department of War in September 2025 by President Donald Trump) AI program called Project Maven.

Pairing AI with Human Decision-Making

Project Maven gathers data from many sources. To understand what it is looking for, it trains on images collected by drones, satellites, and other, nonvisual, surveillance sources such as radar signals, infrared sensors that detect body heat, and geolocation data from smartphones and social media. Then Project Maven AI sifts through this massive amount of input to locate objects of interest, such as aircraft, ships, tanks, or missiles. Once potential targets are identified, the AI system works to determine whether the objects belong to friendly or hostile forces. Human operators then verify whether the AI's recommendations are accurate and decide what action the military should take against the proposed targets.

Using this computer vision AI, the US military was able to specifically identify missiles, rockets, drones, and militia camps that belonged to the enemy in Iraq and Syria. The AI technology also located and identified enemy rocket launchers in Yemen and enemy ships in the Red Sea. In February 2024 the US military sent B-1 bombers and other aircraft armed with precision-guided missiles to conduct eighty-five bomb strikes against some of these targets.

Despite the importance of AI technology in this bombing mission, the US military stressed that human oversight was a vital part of the decision-making process. Schuyler Moore, the US Central Command chief technology officer, explains, "There is never an algorithm that's just running, coming to a conclusion, and then pushing onto the next step."[1]

New Roles for AI

The US military includes the Army, Air Force, Navy, Marine Corps, Coast Guard, and Space Force. All six of these military branches use AI in their work. And AI is proving to be a real asset to the military in both its combat and support roles. AI can help target enemies on a battlefield, but it can also assist in maintaining supply chains that allow troops to operate within that battlefield. Even jobs

Using computer vision AI, the US military can now locate and identify enemy missiles and rocket launchers and precisely target them for destruction.

in the military's health care services, engineering units, intelligence networks, and cybersecurity programs are now augmented by AI. In all these capacities, the military views AI as a teammate to military personnel, with the two working together to complete tasks faster and more efficiently. Speaking at a conference in 2023, US Air Force Chief of Staff General David Allvin said, "I do believe the future is going to be about human–machine teaming. Optimizing the performance and being able to operate at speed."[2]

The changes AI is bringing to the military and the way it manages conflicts are happening at an incredible pace. While they are working to develop groundbreaking AI technology that provides battlefield advantages, the US armed forces grapple with ethical questions surrounding the responsible applications of this powerful technology. Some fear that AI might fail to discriminate between an enemy and a civilian target, make a lethal decision without human guidance, and even prompt governments to choose a push-button war over complex negotiations. Speaking at a conference in 2024, Deputy Secretary of Defense Kathleen Hicks said, "Even as we're swiftly and safely embedding AI into many aspects of our mission—from battlespace awareness, cyber, and reconnaissance, to logistics, force support, and other back-office functions—we do so mindful of AI's potential dangers, which we're determined to avoid. That's why we always have a human responsible for the use of force."[3]

"I do believe the future is going to be about human–machine teaming. Optimizing the performance and being able to operate at speed."[2]

—General David Allvin, US Air Force chief of staff

As AI impacts the way in which the military performs its mission, strategists and soldiers must stay informed about emerging technologies. From combat decision-making to autonomous weapons to intelligence gathering, the military is working faster and more efficiently because of AI. Yet military leaders also recognize that potential adversaries are learning from and deploying these same tools. With the advent of AI tools, keeping the peace will remain an important charge of the armed forces.

CHAPTER ONE

Seeking the AI Advantage

AI is shaping the ways armed forces operate in the modern world. Military applications of AI started in the middle of the twentieth century and continue today as the technology evolves. Different types of AI help military personnel and other types of combatants locate targets, guide weapons with greater precision, accelerate decision-making, and maximize the efficient use of resources.

In general, AI supplies military forces with data needed to make quicker, better-informed decisions. AI technology can analyze huge amounts of data in just seconds. It works much faster than humans can work and completes tasks without the fatigue that humans experience. Therefore, complex jobs that might take humans weeks or months can be achieved by AI in just seconds. This kind of speed allows the military to handle its varied tasks more quickly and efficiently. In a 2023 speech, US Deputy Secretary of Defense Kathleen Hicks explained, "As we focused on integrating AI into our operations responsibly and at speed, our main reason for doing so has been straightforward: because it improves our decision advantage."[4]

> "As we focused on integrating AI into our operations responsibly and at speed, our main reason for doing so has been straightforward: because it improves our decision advantage."[4]
>
> —Kathleen Hicks, former US deputy secretary of defense

The United States is not the only country working to develop AI in wide-ranging industries. Countries like China, France, India, South Korea, the United Arab Emirates, and the United Kingdom and are also prioritizing their AI capabilities. Even insurgent groups and non-state actors are learning how to leverage AI to conduct hostile actions. Staying ahead of enemies

> "Advances in AI have the potential to change the character of warfare for generations to come. Whichever nation harnesses AI first will have a decisive advantage on the battlefield for many, many years."[5]
>
> —Mark T. Esper, former US secretary of defense

in the race to master AI's capabilities might help decide the outcomes of twenty-first-century conflicts and beyond. As US Secretary of Defense Mark T. Esper said in 2019, "Advances in AI have the potential to change the character of warfare for generations to come. Whichever nation harnesses AI first will have a decisive advantage on the battlefield for many, many years."[5]

In 2024 Stanford Institute for Human-Centered AI conducted a survey about AI's influence on society. The survey results showed that the United States leads the world in AI development. This ranking was determined by considering the country's AI research and development, its AI-related economic activity, and its AI infrastructure. The results show that the United States produces the highest-quality AI of any of the thirty-six countries included in the survey. China follows in second place. But the rankings are far from secure as nations compete to gain advantages with a technology that is still not widely understood.

What Is AI?

Human beings think, learn, create, communicate, make decisions, and solve problems in complex ways. AI is technology that imitates many of these aspects of human intelligence. To learn and make decisions, AI technology uses powerful computers that gather and analyze large amounts of data. As the AI system learns from data, its performance improves. Algorithms are the instructions or rules that AI technology uses to perform various tasks. These algorithms allow AI machines to recognize patterns in data, respond to the data, classify it into categories, and make predictions based on it. In the case of generative AI, such as ChatGPT or DALL-E, the algorithms allow the programs to understand and produce natural language or imagery. AI algorithms also improve through experience, learning from larger datasets and human feedback. It is these abilities to analyze, sort, react, and learn that allow AI to perform many tasks in ways similar to humans—but much faster.

AI has become a part of people's everyday lives. Voice assistants like Siri, Alexa, and Google Assistant use AI to answer questions or control devices. Refrigerators equipped with AI can create grocery shopping lists. Streaming services like Netflix, YouTube, and Spotify use AI to suggest movies, programs, or music a person might enjoy. Navigation apps like Google Maps and Apple Maps use AI to give drivers directions and real-time traffic news. Social media platforms like TikTok, Instagram, and Facebook use AI to keep users engaged. Online shopping platforms use AI chatbots to respond to customer service questions. Search engines like Google, Yahoo, and Bing use AI to respond to users' questions.

These routine uses of AI are relatively recent. In contrast, the US military has worked to harness the power of AI since the 1950s. One of the earliest uses involved computer-assisted pattern recognition to help radar operators track aircraft and missiles. Since then, the military has found ways to apply much more powerful AI technology to its mission. Today AI handles the vast amounts of data that are gathered by satellites, aircraft, radar, drones, and other communications resources to coordinate battlefield command and logistics. Sam J. Tangredi, a retired US Navy captain, said, "Investments in

AI is shaping the ways armed forces operate in the modern world. Military applications of AI started in the middle of the twentieth century and continue today as technology evolves.

AI are justified, because the U.S. armed forces are awash in operational and tactical information. Data appears everywhere. A better means of compiling, storing, and categorizing information *very rapidly* is essential to make sense of the ever-expanding amount of data warfighters are collecting through digital systems."[6]

Traditional AI and Generative AI

The military uses both traditional AI and generative AI. Traditional AI works with programmed instructions to sort and categorize data to produce a response. Amazon employs traditional AI to suggest products to customers based on their purchasing history. This kind of AI is very rigid in its application of rules, but it is excellent at recognizing patterns. It is typically precise and reliable, and the military uses it to identify targets, create terrain maps of battlefields, and program unmanned aircraft and guided missiles to reach their destinations.

Generative AI learns patterns and rules and then produces new content based on those factors. Depending on a user's requests,

Machine Learning and Deep Learning

Generative AI is built on machine learning, particularly a subset known as deep learning. Machine learning describes AI systems that learn from historical data. In the 1950s Arthur Samuel, an early leader in the AI field, defined machine learning as "the field of study that gives computers the ability to learn without explicitly being programmed."

Machine learning is widely used in computer systems today. It plays an important role in recommending social media posts, predicting diseases in health care, and suggesting products a user might be interested in. For the military, machine learning is useful in a wide range of tasks, such as protecting computer networks, guiding robotic systems, and identifying targets.

Deep learning describes machine learning models that imitate the way the human brain works. The US military is using deep learning in a statistical model that can learn features or elements in images without human guidance. For example, the computer can identify a building or vehicle in an image. With the deep learning technology, this model is much more effective than current models for identifying objects in low-resolution photographs, providing analysts with better knowledge of enemy troop movements, resource depots, and gatherings of military hardware.

Quoted in Sara Brown, "Machine Learning, Explained," MIT Sloan School of Management, April 21, 2021. https://mitsloan.mit.edu.

this content can be text, images, video, music, or coding. For example, AIVA is an AI tool that allows musicians and songwriters to generate music in various styles because the program has trained on many different musical types.

The military is also experimenting with generative AI. In 2024 the US Army was the first branch to bring a generative AI platform into its operations. The Ask Sage AI platform allows the Army to increase its speed in completing a variety of tasks. These tasks range from writing computer code to analyzing documents, and Ask Sage performs these tasks with fewer mistakes than humans make. Using Ask Sage, military personnel can write computer code thirty-five times faster than when they are working without the AI platform. Analyzing thousands of documents used to require days of work for Army clerks. Now personnel can analyze the same number of documents in just minutes. This has improved cybersecurity, intelligence gathering, logistics planning, and several other critical operations that permit military forces to function more efficiently.

Precursors to Modern Military AI

It has taken decades to develop modern AI technologies. In the 1940s researchers began developing electromechanical computing machines to accelerate tasks that had traditionally required human reasoning. Alan Turing, a British mathematician, made major contributions to the field of code breaking by designing a computing machine that could test thousands of code possibilities in a short amount of time. Famously, the device he and his team built—known as the Bombe—was instrumental in deciphering Germany's complex Enigma code during World War II. The Enigma's settings changed daily, but once its key parameters were determined, the Bombe could search through tens of thousands of possible combinations in roughly twenty minutes to identify the correct configuration, enabling Allied cryptanalysts to read intercepted messages. Today modern computers and AI algorithms can replicate the Enigma decryption process in mere seconds—a testament to how far computing technology has advanced. In 1950 Turing wrote a paper titled "Computing Machinery and Intelligence," in which he questioned whether machines could think. Throughout the paper, Turing proposed ways to assess whether machines could exhibit humanlike intelligence and even learn from experience. Turing's ideas remain foundational for AI research today.

From the 1940s through the 1980s, scientists and scholars studied ways to apply this new technology in military and civilian contexts. The US military for example, focused its research efforts by forming the Defense Advanced Research Projects Agency (DARPA) in 1958. DARPA remains the US Department of Defense's research and development agency. Its mission, as stated on the DARPA website, is "to create technological surprise for U.S. national security."[7] From its inception, DARPA has supported research on AI technologies. As of 2024 about 70 percent of DARPA's research projects incorporated AI technologies.

In the 1990s the military shifted its focus from AI research to AI applications. The military's first practical use of AI was the Dynamic Analysis and Replanning Tool (DART), employed during the Persian Gulf War (1990–1991). In this conflict, the US military joined other countries to free Kuwait from an invasion by its neighbor, Iraq. Using DART, the military created highly efficient transportation schedules for troops and supplies. DART also helped military commanders' decision-making by evaluating various options for moving troops and supplies and identifying the most economical, efficient plans in minutes. With the scheduling and decision-making that DART AI technology provided, the US military saved millions of dollars during the Gulf War. DART demonstrated how successful AI technologies could help the military do its many complex jobs.

Self-Driving Vehicles and Neural Networks

Over time, the US military looked at ways that AI could help with tasks that are even more complex than scheduling and simple decision-making. Innovators explored how AI could be used in autonomous vehicles, on land, at sea, and in the air. In 2016 the

Navigation apps like Google Maps and Apple Maps use AI to give drivers directions and real-time traffic news.

US military launched the autonomous vessel *Sea Hunter*. It was a prototype vessel, built to demonstrate the US military's advanced technologies, including AI. This AI-driven unmanned ship can operate at sea without humans aboard for months at a time. *Sea Hunter* tracks enemy submarines, gathers intelligence, and detects underwater mines. It does these jobs at much lower costs than similar manned ships. Interviewed the year of *Sea Hunter*'s launch, US Deputy Secretary of Defense Robert Work described the vessel, saying, "This is the first time we've ever had a totally robotic, trans-oceanic-capable ship."[8]

In the 2020s the US military is focusing attention on developing more sophisticated neural networks. This deep learning technology allows AI to perform better than humans in tasks like image recognition and natural language processing. Such AI technology has been used in Ukraine to stem Russia's invasion that began in 2022. There, AI neural network tools help the Ukrainian military translate and analyze enemy communications much more quickly, providing more preparation time to counter Russian plans.

DARPA Challenge Competitions

In 2004 the Defense Advanced Research Projects Agency (DARPA) hosted the Grand Challenge race for robotic, self-driving cars. The fifteen competitors attempted to race their entries over a 142-mile (229 km) course in California and Nevada. DARPA hosted the competition to encourage development of autonomous vehicles. Its goal was for the military to someday use autonomous vehicles to move supplies to combat areas, removing human drivers from those dangerous jobs.

DARPA offered a $1 million prize for the first team to finish the race in under ten hours. However, no cars completed the rough, desert course. DARPA still called that first challenge race a success. Lieutenant Colonel Scott Wadle, DARPA's liaison to the US Marine Corps at that time, said:

> That first competition created a community of innovators, engineers, students, programmers, off-road racers, backyard mechanics, inventors and dreamers who came together to make history by trying to solve a tough technical problem. The fresh thinking they brought was the spark that has triggered major advances in the development of autonomous robotic ground vehicle technology in the years since.

Since that first Grand Challenge, DARPA has sponsored competitions in fields like cybersecurity, AI, and robotics with a goal of fueling new technologies for military and civilian applications.

Quoted in Denise Chow, "DARPA and Drone Cars: How the US Military Spawned Self-Driving Car Revolution," Live Science, March 21, 2014. https://www.livescience.com.

Researchers are making new advances in AI technology all the time. Both in the military and in civilian industries, keeping up with the evolving technology is crucial. At a 2024 press briefing, US National Security Advisor Jake Sullivan said, "A specific AI application that we're trying to solve for today in the intelligence or military or commercial domains could look fundamentally different six weeks from now, let alone six months from now, or a year from now, or six years from now. The speed of change in this area is breathtaking."[9]

Ethical Questions

Over time, the US military has pursued ways for AI to automate routine tasks, allowing personnel to work on more complex duties. AI can improve decision-making speed and efficiency, especially amid the chaos of the battlefield. Precision guidance of military hardware is aided by AI, and entirely unmanned weapons and vehicles help keep soldiers from risking their lives. AI is also available around the clock and does not make fatigue-related errors like humans do. But there are also ethical concerns that accompany AI technology use in combat.

AI technologies often make decisions and act in ways that resemble human reasoning and behavior. In wartime, military commanders make decisions that put the lives of their soldiers in danger. They also can show restraint, knowing when it is appropriate to attack an enemy and when it might be better to show mercy. Similarly, commanders must consider collateral damage or civilian casualties that might result from their actions. One ethical concern, then, is how much AI technologies should assume such responsibilities.

AI systems can easily find and identify an enemy target. That target might be a ship, a plane, a large weapon system, or an enemy encampment. Once a target is identified, the AI system

The Navy's unmanned surface vessel Sea Hunter *transits underneath the Sydney Harbor bridge in Australia. This AI-driven ship can operate at sea without humans aboard.*

can aim and fire a weapon. The automated response in such a scenario requires no human input, and that worries critics.

The US military, however, continues to stress that it maintains boundaries on AI use so that humans are the ultimate decision-makers concerning the use of deadly force. Speaking in 2023, Kathleen Hicks clearly stated, "Our policy for autonomy in weapon systems is clear and well-established: There is always a human responsible for the use of force."[10] For example, the US military's Project Maven is capable of complex target finding. But once enemy targets are found, humans always confirm the AI's targets and make decisions on whether a deadly force response is necessary.

Even though humans are involved in the decision-making process, many fear that AI dehumanizes conflict. AI-guided munitions can deliver payloads well behind enemy lines, causing death and destruction to both military and civilian populations. Some observers focus on the impressive capabilities of the technology, overlooking the real human cost of its use. As the Russian invasion of Ukraine has shown, soldiers near the battlefield are increasingly vulnerable to AI-assisted drone strikes, which have made combat deadlier and less predictable.

Alongside the concern that AI is dehumanizing conflict is the question of whether AI might also escalate it. Because AI can analyze information and make decisions much faster than humans, wars might escalate simply because of AI's speed. The immediate response capabilities could allow nations to choose to take hostile actions without pursuing diplomacy. Similarly, because AI-augmented weapon strikes often seem as if they are "risk-free" to an attacker, leaders might consider them as a fast and inexpensive way to justify warfare. The United States and other countries continue to discuss these ethical concerns and how they might shape future global relations.

AI in Battle

In the chaos, confusion, and uncertainty of the battlefield, military commanders must make decisions at an extremely fast pace. Since these decisions might shorten a battle and reduce casualties, it is beneficial to have the most recent reports possible about enemy positions and capabilities. AI technology can process vast amounts of information about the enemy's strength and location, enabling commanders to speed their decision-making process. But this is only one aspect of AI use in the modern military. AI can help guide missiles or bombs to targets; it can also aid logistics by keeping track of supplies and directing troops to where they are needed.

AI in Command and Control Systems

During an operation, military commanders assess the battlefield situation, make decisions, and give personnel combat orders. The overall planning and execution of an operation is known as command and control, the coordination of military forces and their supplies to accomplish the mission. Speaking at a conference in 2023, Deputy Secretary of Defense Kathleen Hicks said, "From the standpoint of deterring and defending against aggression, AI-enabled systems can help accelerate the speed of commanders' decisions and improve the quality and accuracy of those decisions, which can be decisive in deterring a fight and winning in a fight."[11]

"From the standpoint of deterring and defending against aggression, AI-enabled systems can help accelerate the speed of commanders' decisions and improve the quality and accuracy of those decisions, which can be decisive in deterring a fight and winning in a fight."[11]

—Kathleen Hicks, former US deputy secretary of defense

The US military's Project Maven is an AI program that is helping with command and control tasks. Military commanders have always needed to know where their enemy is located. In the early 1990s, when the Gulf War was fought, Project Maven had not yet been developed. At that time, the US military used the Global Positioning System to track and target the enemy. Human operators analyzed the data collected. This technology allowed them to identify enemy targets and take action against the enemy within a few hours. The technology continued to improve so that by the early 2000s, the US military could find and identify enemy targets and attack them within ten to fifteen minutes. However, that speedy response time required more than one thousand military personnel working at the same time to process the data to get information to the commanders quickly.

Project Maven's ability to gather and analyze visual and other input to identify potential battlefield threats is an example of the application of AI to US command and control operations. The US military has tested Project Maven through a series of military exercises

The US Army conducts a disaster relief coordination meeting as part of a multinational exercise. During an operation, military commanders assess, make decisions, and give orders.

named Scarlet Dragon. In 2020 the military used Project Maven at Fort Liberty in North Carolina, in a Scarlet Dragon training exercise designed to test the system's capabilities. During the training, the Project Maven system—trained on images of vehicle types—correctly identified an enemy tank that was part of the exercise and suggested it as a possible target. The operators confirmed the target and sent the message to fire a rocket. The rocket hit and destroyed the tank. It was the first time ever that American soldiers had attacked a target that was located and identified by AI. Army Colonel Joseph O'Callaghan, who led the AI targeting for that training exercise, says, "It showed us the art of the possible."[12]

Project Maven collects data from a variety of sources, such as drones and satellites, sensors, and geolocation information. This geolocation data comes from smartphone activity, social media tags, and IP addresses. Project Maven processes this massive amount of data using machine learning algorithms. Then the AI algorithms identify objects within the data that are of interest to the military.

Project Maven is highly accurate in identifying targets, such as trucks, tanks, ships, aircraft, and rocket launchers. The system is also able to distinguish between enemy targets and friendly military forces. Once Project Maven identifies the target, it tracks the target, if it is moving, and provides military commanders with strategies for attacking the target.

Automatic Target Recognition

Project Maven uses automatic target recognition, an AI technology that employs sensors and machine learning algorithms to identify targets. As the technology continues to improve, Project Maven could be capable of allowing small military teams to strike up to one thousand targets in an hour. Clifford A. Baxter, a US Army warrant officer, explains, "AI-driven systems can significantly enhance precision and accuracy in tactical targeting. Traditional targeting methods often rely on human operators, who can be prone to fatigue, stress and human error. AI algorithms, on the other hand, can analyze vast amounts of data quickly and accurately, leading to improved target identification and tracking."[13]

Automatic target recognition uses a variety of sensors on aircraft, satellites, and drones to gather data. That data is processed, and the system runs algorithms to find features on the object it has detected. These features include things like the shape and structure of the object and its motions. Then the system uses machine learning to compare the features it detected against a huge database of known targets. The automatic target recognition system can accurately identify that object within seconds. After the system identifies the object, the human operator examines the object to confirm that the AI is correct. "AI-driven targeting systems can revolutionize precision, accuracy and sensor-to-shooter capabilities, elevating the effectiveness and efficiency of military engagements to unprecedented heights,"[14] says Baxter.

Since those first training exercises at Fort Liberty, the US military has used Project Maven in actual military operations. Project Maven has helped find enemy targets on land in Yemen, Iraq, and

AI-driven automatic target recognition uses a variety of sensors on satellites (pictured) and drones to gather data.

AI Helps Soldiers "See" the Battlefield

The US Army is developing a headset that uses AI to help soldiers make better decisions during training and on the battlefield. The Integrated Visual Augmentation System (IVAS) headset is a pair of goggles that soldiers wear in battle. IVAS uses AI technologies to combine data from sensors, unmanned aircraft, and other platforms and then provides critical information the soldier needs on the headset display.

IVAS gives soldiers an enhanced awareness of their environment that is not possible with human eyesight alone. For example, with IVAS, soldiers can see what is outside of their armored transport before they leave the safety of its interior. IVAS brings video images from drones or vehicle-mounted cameras to the soldier's headset. Soldiers can operate safely when AI helps them become more aware of their environment. IVAS is equipped with low-light, thermal night vision so soldiers can even "see" through smoke or in darkness.

As of 2025 the US Army was testing IVAS and gathering feedback from soldiers on its effectiveness. Based on the feedback, the Army continue to refine the technology.

Syria. It has also found hostile ships in the Red Sea. The United States is not alone in employing AI to find battlefield targets. In recent conflicts, both Israel's and Ukraine's military have used AI technology for the same purpose.

AI in Strategic Support Systems

Strategic support systems or decision support systems are other arenas in which AI is helping the military achieve its goals in combat. While command and control systems in the military are focused on carrying out a specific mission, the strategic support systems are focused on longer-term planning and decision-making. With the help of AI, military data collection is greatly accelerated. AI can speedily recognize connections and patterns that would take much longer for humans to identify and analyze. It can also analyze large amounts of data and give military personnel a more complete picture of a hostile environment in seconds, helping leaders make better decisions.

US Navy researchers have developed a decision support system called LittoralLens. The system uses computer vision technology to assess environmental data along coastlines in real time. With AI computer vision technology, machines can see and understand the visual world from images and videos. For example, LittoralLens can compute information about waves and surf strength in real time, assisting sailors and marines who are conducting operations along a shoreline. The portable system can be operated by one person working alone and is an uncomplicated "point-and-shoot" tool. Without AI, it can take the military up to forty-five minutes to collect surf observations. LittoralLens shortens that assessment period and provides personnel with crucial information that might influence the success of an amphibious operation.

AI in Battlefield Casualty Care

AI assists in battle planning and target finding, but it also helps save wounded soldiers. When a soldier is injured in battle, a battlefield medic is often the first to administer care. Initially, a medic will fill out a tactical combat casualty care card, or TCCC card, for the injured person. This card describes the soldier's injuries, the medical treatment given on the battlefield, and potential evacuation information if the wounds are severe. This TCCC card is essential for the soldier's future care and recovery.

PATRIOT Missile System Tracks Targets

The US Army's Phased Array Tracking Radar to Intercept On Target (PATRIOT) air defense missile system uses AI to track targets. The PATRIOT system utilizes a type of AI called deep learning to monitor incoming objects and determine whether they are enemy or friendly objects. These incoming objects could be tactical ballistic missiles, cruise missiles, or advanced aircraft.

With deep learning, the computer systems process large amounts of data to identify relationships and patterns that humans cannot easily see. Thus, PATRIOT can track multiple incoming objects at the same time and can operate in harsh environments when vision is poor. Once a target is acquired, the PATRIOT system can launch surface-to-air missiles in response.

The PATRIOT was first used in combat during the Gulf War in 1991. In that conflict, the US military joined other countries in fighting against Iraq when Iraq invaded Kuwait. The US military has continued to make the PATRIOT system more autonomous in the years since then. In 2022 the US military began providing PATRIOT missile systems to Ukraine to aid in stemming the Russian invasion of that nation.

However, in the chaos of the battlefield, a wounded soldier's TCCC card might become lost or damaged, never reaching the hospital or aid station personnel who will provide more extensive treatment. Researchers interviewed senior combat medics who said that only 10 to 15 percent of the TCCC cards arrive with patients sent to hospitals, and that can limit a seriously wounded soldier's chance of survival. Addressing the importance of speed and accuracy of care, the late military surgeon R. Adams Cowley said, "There is a golden hour between life and death. If you are critically injured, you have less than 60 minutes to survive."[15] Others insist the survival window is even shorter. Just filling out the TCCC card takes up some of those precious minutes.

Researchers are developing AI technology that fills out TCCC cards automatically as it processes data from sensors as well as audio and video collected as the medic treats the patient. The sensors collect the patient's vitals, such as heart rate, blood pressure, and oxygen levels. As the AI processes this data, it fills out the TCCC card. Then the card is sent to the hospital staff in digital form. The hospital staff gets more complete information on the incoming patient, allowing them time to better prepare for the necessary care. AI saves time on the routine administrative parts of completing the TCCC card on the battlefield, allowing the medics to focus more on caring for the patient. Ultimately, researchers believe that this speedy, coordinated process will save soldiers' lives.

Other Medical Care AI Tools

AI is aiding military health care in other ways, too. In the United Kingdom researchers are developing an AI-powered robotic drone that helps military medics on the battlefield. During the Vietnam War, American troops used helicopters to quickly evacuate wounded soldiers from the battlefield to the hospital. Evacuating wounded soldiers by air has worked well in the years since

During the Vietnam War (pictured), American troops used helicopters to quickly evacuate wounded soldiers from the battlefield to the hospital.

the Vietnam War. However, with the development of shoulder-launched missiles on the battlefield, it is more dangerous for helicopters to reach evacuation sites. Other ways to help stabilize patients on the battlefield were needed if airlift evacuation was too risky.

ATRACT stands for A Trustworthy Robotic Autonomous system to support Casualty Triage. The ATRACT system uses a robotic autonomous system drone that helps medics on the battlefield make quick decisions in those important first few moments after a soldier is wounded. The ATRACT drone uses visual and thermal imaging data to search for and find wounded soldiers in difficult terrain. The drone then uses its sensors to determine the soldier's vital signs and assesses how severe the injury is. It relays this medical information to the medics while they are en route to reach the wounded soldier. Where there are multiple wounded soldiers in an area, ATRACT helps the medics determine which soldiers are most severely injured and need care first. "There is an urgent unmet need for enhancing casualty survival in a warzone where conventional helicopter evacuations simply aren't possible. While there is much talk of using AI and autonomous platforms as weapons, this project does the opposite, helping to save lives, protect soldiers, and make war less deadly,"[16] says Ardhendu Behera, professor of computer vision and AI at Edge Hill University in England, where ATRACT was developed.

On the battlefield, where decisions must be made quickly, AI is proving beneficial. From locating and targeting potential threats to helping wounded soldiers receive lifesaving medical care, AI technologies are changing the logistics and the conduct of war in the twenty-first century.

AI in Aircraft, Weapons, and Ships

The military has used motorized vehicles and weapons in combat for more than one hundred years. Today the military is using AI in its vehicles and weapon systems to give them autonomous capabilities that allow them to operate with minimal support from military personnel. AI allows these vehicles and weapon systems to identify and track targets and helps military commanders make rapid decisions on the battlefield.

Unmanned Aircraft Testing Grounds in Ukraine

One of the more prominent uses of AI is to guide unmanned aerial vehicles, commonly called drones. The guidance is provided by an unmanned aerial system that includes the ground control station and communications system needed to monitor the drones and direct them to targets. With AI, the unmanned aerial system can process data quickly, analyze complicated environments, and make decisions.

Small quadcopter drones have played a very important role for both sides in Russia's war against Ukraine that began in 2022. These drones are called quadcopter drones because they have four propellers. In June 2025 the Ukrainian military launched Operation Spiderweb against Russian military infrastructure. Over one hundred Ukrainian quadcopter drones attacked air bases in key locations deep inside Russia. Their targets were the Russian military's long-range bomber aircraft designed to carry nuclear weapons.

The drones were smuggled in pieces into Russia, where they were reassembled by Ukrainian patriots with access to trucking fleets. The drones were then hidden in cabinet-like structures commonly seen on flatbed trucks. Unsuspecting Russian drivers were hired to move the trucks and park them near Russian air bases. Then a remote command was given, and the cabinet roofs opened, allowing squads of drones to launch. Each of the drones had an individual pilot back in Ukraine controlling its flight. But the drones were designed to use onboard AI technology for navigation if jamming occurred or if the signal between the operator and the drone was disrupted by the vast distances. The drone's AI technology had also been trained to recognize specific models of Russian aircraft and could identify and attack vulnerable places on the aircraft, like fuel tanks and weapons. The Ukrainian military claimed success when its drones hit forty-one Russian aircraft and destroyed at least thirteen of them in Operation Spiderweb. "Of course, not everything can be revealed at this moment, but these are Ukrainian actions that will undoubtedly be in history books,"[17] said Ukrainian President Volodymyr Zelenskyy in June 2025.

Quadcopter drones like those used in Operation Spiderweb are relatively inexpensive, costing about $600 to $1,000 each. The estimated but unverified cost of the Russian aircraft destroyed in the attack was $7 billion, according to Ukraine.

The US Military Learns to Adapt

The war in Ukraine and the success of missions like Operation Spiderweb are demonstrating the value of drones in modern combat. While many major military powers have traditionally amassed fleets of expensive aircraft, warships, and land vehicles, the focus is shifting to cheaper options that can be as effective. In 2025 the US military began training new units to use small drones for reconnaissance and combat. For instance, military units are building and 3-D-printing drones on the battlefield for quick deployment. Like the drones used in Ukraine, these unmanned aerial vehicles can employ AI to navigate when human-controlled systems are jammed, allowing the drones to reach their targets and identify

Small quadcopter drones like this one being used by the Ukrainian military have played a very important role for both sides in Russia's war against Ukraine, which began in 2022.

the best strike points to ensure destruction. "The first fight of the next war is going to involve more drones than any of us have even seen,"[18] says US Major General Curt Taylor.

US soldiers training at Fort Bliss in Texas know how to build their own drones so that they can modify or repair them in the field. The soldiers learn how to fly the drone, using a simulator that works something like a video game controller. By practicing on the simulator, they gain flight experience. The US military's goal is for soldiers to think of the drone "as if it was their personal weapon, their radio, their night-vision goggles or a grenade," said Colonel Nick Ryan in 2025. "That it's just something they're so used to and so familiar with, that it's just part of their standard kit that they take with them everywhere they go."[19]

> "The first fight of the next war is going to involve more drones than any of us have even seen."[18]
>
> —Curt Taylor, US major general

AI is also being adopted in US military aircraft in Project VENOM. The Air Force is modifying several F-16 fighter jets so AI can pilot the aircraft. Currently, human pilots are accompanying the AI to test its features, responsiveness, and capabilities, but the goal is to use these older-model fighter planes as autonomous "wingmen" that will fly alongside crewed aircraft. "Modifying the aircraft is the result of a rigorous design phase and brings us one step closer to testing autonomy on a fighter jet with real mission systems and capabilities,"[20] says Major Trent McMullen with the 40th Flight Test Squadron at Eglin Air Force Base in Florida.

Project VENOM simulators are running tests to put the AI through combat scenarios. "These simulations provide an efficient way to train the autonomy to learn complex air combat tactics." McMullen says. "As the VENOM program's first flight approaches, we are excited to test novel autonomous solutions. The strides we've witnessed in the simulation environment suggest VENOM (Viper Experimentation and Next-gen Operations Model-Autonomy Flying Testbed) will help advance aerial combat capabilities for future crewed and uncrewed platforms."[21]

AI in Weapon Systems

In addition to placing AI in its aircraft, the military is also using AI in its weapon systems. These autonomous systems include missiles and rockets that use AI for navigation and target recognition. The Aegis Combat System is an example of older technology the

AI Tools Loaded to Soldiers' Phones

US Army soldiers may soon discover that the AI tools they need for their job are as close as their phone. The Army is installing AI tools on soldiers' phones, laptops, and drones that can help them detect enemy threats. TurbineOne is the name of an AI-powered app installed in soldiers' gear in 2025. US Army troops have tested the app in training exercises in Europe and in the Pacific.

Using the app, soldiers can quickly locate an enemy's position or their drone launch sites based on AI processing signals from aircraft, reconnaissance drones, and other sources. In the past, soldiers depended on analysts who sorted through massive amounts of data collected from drones, planes, and satellites for this type of information. Most importantly, the app provides this information to soldiers in the field without requiring an internet connection, which might be jammed or could alert enemy troops to the user's location. TurbineOne's system is allowing individual soldiers to harness AI to help them operate effectively on the battlefield.

US military has updated to include AI. The US Navy has used the Aegis system for decades. The Navy commissioned its first Aegis ship in 1983. By 2021, Aegis systems had been installed on eighty-four US Navy ships.

The Aegis system integrates computers, sensors, and radar to detect and track enemy targets, assess threats, and direct weapons to intercept them. Aegis can simultaneously track multiple enemy threats from the air, surface, and underwater. In recent years AI has been added to the Aegis system to improve its radar capabilities and its ability to detect enemy threats. The AI quickly processes tremendous amounts of complex data to identify incoming threats and provide military commanders with response options in seconds. AI also predicts when parts of the Aegis system might need maintenance. This keeps the Aegis systems running efficiently and reduces repair costs.

The US Navy's destroyers patrolling the Red Sea have Aegis systems on board. There, they use Aegis to defend against missile and drone attacks from Houthi rebels who have been disrupting international shipping in the region. In 2023 the US Navy and its allies began responding to Houthi hostilities. The Houthis attack using unmanned systems under the water, on its surface, and in the air. The Aegis systems react and direct US countermeasures. The real-time data is incorporated by the AI, permitting it to learn on the spot and improve the overall system.

Concerns over Lethal Autonomous Weapons

While AI advances the capabilities of weapon systems, there are some ethical questions surrounding its use. Lethal autonomous weapon systems (LAWS) refer to weapons that can operate independently of human input. These weapon systems are characterized by the exclusion of humans from the decision-making process and are often informally referred to as "killer robots" because of their high degree of autonomy.

The Air Force is modifying several F-16 fighter jets so AI can pilot the aircraft. The goal is to use these older-model fighter planes as autonomous wingmen to support crewed aircraft.

LAWS use sensors and computer algorithms to identify an enemy target and then launch a weapon against that target. A report issued by the US Congress in early 2025 stated that the United States does not currently have any of these weapons in its weapons inventory. US military leaders have said that if other countries pursue these types of weapons, then the United States may feel the need to develop their own. Many international groups are calling for LAWS to be banned because of concerns over having a weapon system that operates without human supervision.

The ongoing ethical question that many militaries around the world wrestle with is how to incorporate AI without crossing such ethical boundaries. In the Israel–Gaza conflict that began in 2023, Israel has used AI for decision support. The AI technologies create lists of people and buildings that are possible targets. Then a human decides whether to use the weapon system to strike that target.

AI-Powered Robot Dogs

The US Army has been using dogs for military work since World War I. In that war, dogs carried messages through the trenches. In World War II, the US Army trained dogs to serve on sentry duty, perform scout or patrol work, carry messages, and detect mines.

Today dogs are still doing military work—but some modern canine companions are robots. In 2021 the first semiautonomous robot dogs arrived at Tyndall Air Force Base in Florida. The official name of these robot dogs is Quad-legged Unmanned Ground Vehicles (Q-UGVs). The robot dogs can navigate in rough terrain, and at Tyndall they are used to patrol remote areas of the military base.

In 2024 the Army tested the Q-UGVs during military exercises in Saudia Arabia. In this test, a robot dog was armed with an automatic rifle operated by the onboard weapon system. Its programming uses artificial intelligence for target finding. The robot dog fired on static ground targets, but it is hoped that the weapon might be used to destroy airborne drones.

Robot dogs are currently being used in combat in the Russo-Ukrainian War. The Ukrainian army uses robot dogs to deliver ammunition to troops and to conduct surveillance.

Critics argue that military personnel may depend too heavily on the AI's choice of targets without fully considering all the factors, such as collateral damage and civilian deaths, involved in sending weapons to make a lethal strike. Other critics raise concerns that the AI may incorrectly identify a target, leading to wrongful death.

> "There is no place for lethal autonomous weapon systems in our world. Machines that have the power and discretion to take human lives without human control should be prohibited by international law."[22]
>
> —António Guterres, United Nations secretary-general

In May 2025, United Nations Secretary-General António Guterres called for a global ban on lethal autonomous weapon systems. "There is no place for lethal autonomous weapon systems in our world. Machines that have the power and discretion to take human lives without human control should be prohibited by international law," Guterres said. The discussions at the United Nations focused on weapons, like drones, that can find an enemy target and attack it without any human involvement. "Human control over the use of force is essential. We cannot delegate life-or-death decisions to machines,"[22] stressed Guterres. While AI technologies advance in complexity and capability, the debate over the degree to which AI should be used in launching weapons continues.

AI in Unmanned Ships

Along with its aircraft and weapon systems, the US military is using AI to guide unmanned ships. Equipped with AI technology, these unmanned vessels handle a variety of military operations in the water.

The US Navy's USX-1 *Defiant* is an unmanned ship launched in 2025. The *Defiant* is part of the Navy's No Manning Required Ship (NOMARS) program. Unlike other Navy ships, the NOMARS program goal is to design a ship that operates with no humans on board. The Navy has other unmanned ships, but these are manned vessels that have been converted to operate autonomously. The *Defiant* was designed from the beginning to be an unmanned ship.

The *Defiant* uses AI technologies for its autonomous operation. With AI, the ship can safely navigate and avoid collisions. AI

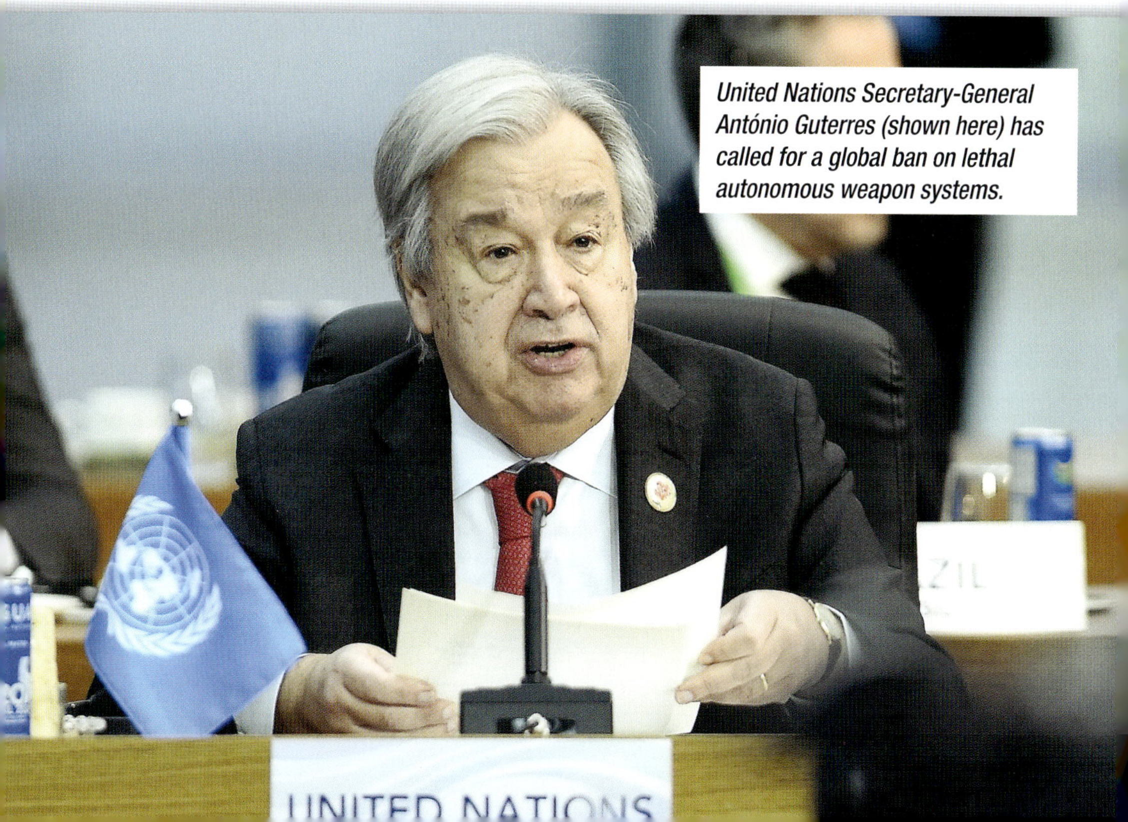

United Nations Secretary-General António Guterres (shown here) has called for a global ban on lethal autonomous weapon systems.

controls its propulsion system that allows the ship to steer and move safely through the water. Its sensors use AI to detect any kind of threats on the water, in the air, or underwater. "*Defiant* is a tough little ship and defies the idea that we cannot make a ship that can operate in the challenging environment of the open ocean without people to operate her,"[23] says Greg Avicola, a DARPA program manager. The ship can also be used as a platform to house and launch drone swarms. The *Defiant* can sail for months at a time without the need to dock. The 180-foot (55 m) ship can be outfitted with weapon systems and is built to be reliable and efficient. The *Defiant* is designed to move and operate safely in rough seas with 13-foot (4 m) waves but can stay afloat even in 30-foot (9.1 m) waves.

The engineers who designed the *Defiant* also worked toward saving costs in their design. "One of the historic problems with unmanned systems is, they promise to keep sailors or pilots safe, but they don't tend to be less expensive than their manned counterparts. We really worked to change the philosophy and operating principles so that this is much less expensive to do the same mission as a manned platform,"[24] says Ryan Maatta, a marine engineer manager working with NOMARS.

The Navy christened the *Defiant* in August 2025 in Everett, Washington. The unmanned ship headed into the Pacific Ocean bound for Port Hueneme, California. Port Hueneme is the center for the US Navy's unmanned operations. This first journey was the *Defiant*'s at-sea demonstration of her capabilities, and the ship successfully reached port in September.

Today's military is using AI in its aircraft, weapons, and ships. These AI technologies give the weapons and vehicles previously unheard-of capabilities, and they give the military that possesses them a decided advantage on the battlefield.

CHAPTER FOUR

AI in Intelligence Gathering

Intelligence gathering is the process the US military uses to collect and analyze information about its enemies. Information about potential and actual adversaries has always been essential for military strategists and decision-makers. What has changed over the years is the volume of intelligence militaries now have available.

Modern militaries gather such massive amounts of information from numerous sources that they can no longer rely on human analysts alone to sort through and understand all the data. As in other countries, the US military has turned to AI to collect, process, and analyze much of the intelligence it collects. AI can process this intelligence faster than any human can work. This allows military commanders to make informed decisions much faster during wartime.

AI in Drone Swarms

One tool the military uses to gather information is the unmanned drone. Sometimes flying alone and sometimes working in small groups, or swarms, of over one thousand drones, these devices use cameras and other sensors to relay information to their human operators. They may fly over battlefields searching for enemy locations, or they may patrol areas to ensure allied units are not caught by surprise.

When drones are working in a swarm, they use AI to communicate and share data with each other, allowing them to make decisions in real time, navigate the environment safely, and never collide with each other. These affordable and versatile drones

One tool the military uses to gather information is the drone. When drones are working in a swarm (pictured), they use AI to communicate and share data with each other.

combine AI and machine learning to accomplish their movements and their tasks. AI makes the drone swarm so efficient because it processes data so quickly that the swarm can react swiftly to changes or threats in its environment.

The US military uses drone swarms in different ways. When the Army is hauling supplies in vehicle convoys, drone swarms can provide security. These drone swarms can fly in front of and alongside the ground vehicles to detect any possible enemy threats to the convoy. The drone swarm relays live images and data to the convoy commander, who then decides how to respond to the threat. Because these military vehicle convoys can sometimes stretch out a mile or longer, they cannot easily turn around if there is danger ahead. The drone swarm provides the early warning that is essential for the convoy commander to keep the troops, vehicles, and supplies safe.

Drone swarms equipped with cameras and infrared detection can also provide security around the perimeter of a military base.

Soldiers on guard duty use binoculars or night vision devices to look for enemy threats. But they are limited in how far they can see from their position. The drone swarms can constantly sweep the area for threats, providing much wider coverage and returning to the base to recharge when needed.

Sweden is one country working to strengthen its drone swarm capability. Sweden joined the North Atlantic Treaty Organization (NATO) in 2024. NATO was formed after World War II as an international security alliance against the growing might of the Soviet Union. It has grown from twelve countries to thirty-two countries today; Sweden is its newest member. In 2025 the Swedish military launched its drone-swarming program. A single Swedish soldier can control up to one hundred drones at a time. The Swedish military tested its new drone-swarm capability at the Arctic Strike exercise in March 2025. The Swedes use these small, affordable drones for a variety of tasks. "With these drone swarms it is possible to conduct both reconnaissance, positioning, and identification,"[25] says Pål Jonson, the Swedish defense minister.

AI in Physical Security

Beyond drone swarms, AI is used in other forms of intelligence gathering. In Greek and Roman mythology, there exists a frightening sea monster named Scylla that guards the channel waters between Sicily and mainland Italy. Mirroring the legend, the US military adopted the name for a new AI technology used to monitor bases and warn of approaching intruders.

The US military uses Scylla AI, for example, to provide physical security for the soldiers at Blue Grass Army Depot in Kentucky. Working with AI and computer vision technology, Scylla analyzes the video feeds from the depot's security cameras. Scylla does not replace the existing security cameras that have been providing security at the base. Rather, it is combined with the existing system to detect threats. "From the standpoint of deterring and defending

"From the standpoint of deterring and defending against aggression, AI-enabled systems can help accelerate the speed of commanders' decisions and improve the quality and accuracy of those decisions."[26]

—Kathleen Hicks, former US deputy secretary of defense

against aggression, AI-enabled systems can help accelerate the speed of commanders' decisions and improve the quality and accuracy of those decisions,"[26] said Deputy Secretary of Defense Kathleen Hicks in a speech in 2023.

Scylla can detect potential intruders even at a great distance and can even identify firearms an intruder might be carrying. While traditional security cameras provide a video recording of any potential threats, Scylla also uses AI to identify the risks and then notify security personnel. During its testing, Scylla was able to detect an "intruder" carrying a gun and climbing a water tower a mile away from the security camera. "If you're the security operator, do you think you could watch 15 cameras at one time . . . and pick out a gun at 1,000 feet? Scylla can,"[27] says Chris Willoughby, the electronic security systems manager at Blue Grass Army Depot.

Scylla knows when to notify security personnel of a threat and when to ignore nonthreats. It does not notify personnel when it detects deer or wild turkey, for example, that are present on the depot property. It can recognize that tree branches swaying in the breeze or a handgun in a holster is not a threat. Drew Walter, deputy assistant secretary of defense for nuclear matters, says:

> Its capacity to learn in real time and reduce nuisance alarms—which fatigue security personnel and inhibit responses to legitimate threats—addresses a long-standing challenge in physical security. Scylla AI leverages any suitable video feed available to monitor, learn and alert in an instant, lessening the operational burden on security personnel. While humans still make the final decisions regarding threat response, AI augments detection capabilities.[28]

Future plans call for testing Scylla at other military locations.

AI in Foreign Language Translation

Sometimes intelligence gathering requires understanding other languages. The US military uses AI for language translation to speed this process. Military machine translation technology uses AI's deep learning and neural networks to translate human speech or text from one language to another. This machine translation is specially designed to handle the terminology that is specific to military operations.

The language translation allows US military personnel to better understand what enemy forces are saying from the intercepted conversations. The military collects intelligence from open sources like news websites and social media. It also intercepts radio or satellite communications. Being able to translate this information in real time aids the military in detecting threats from enemies. AI technology analyzes the information and prioritizes the content that needs attention.

Using AI to Counter Drone Swarms

The US military is continually developing AI technologies that improve its own weapons while defending against those of its adversaries. For example, US Navy researchers are developing laser weapon systems that use AI to automate the target tracking system on warships. Tracking inbound enemies—especially multiple enemy targets—is a challenging task for sailors in the Combat Information Center on a ship. AI, however, allows these personnel to track, identify, and assess each threat and direct weapons to respond. Even a swarm of drones can be dealt with efficiently.

Destroying each drone in a swarm attack would strain conventional weapon systems that use bullets or missiles, so the Navy has paired its AI response tech with laser weapons. Brij Agrawal, a professor at the US Naval Postgraduate School, remarks:

> Defending against one drone isn't a problem. But if there are multiple drones, then sending million-dollar interceptor missiles becomes a very expensive tradeoff because the drones are very cheap. The Navy has several laser weapons systems being developed and tested. Laser weapon systems are cheap to fire but expensive to build. But once it's built, then it can keep on firing, like a few dollars per shot.

Quoted in Dan Linehan, "NPS Develops AI Solution to Automate Drone Defense with High Energy Lasers," US Navy, February 12, 2025. www.navy.mil.

Language translation also allows US troops to better communicate with their allies who may speak other languages. Armed forces from different countries frequently work together. The US military and its allies train and take part in joint exercises, carry out peacekeeping missions together, and attend military academies in each other's countries. When they are working together in these ways, they must be able to communicate and easily share information. However, human translators are not always available. Machine translation can fill the gaps and provide the communication that is vital to successful military operations. The machine translation technology can even be embedded into an existing instant messaging system so that US troops and their allies can communicate in real time when training or working together. For example, a US Army soldier and a South Korean soldier working together in a joint exercise could send

AI language translation allows US military personnel to better understand what enemy forces are saying in intercepted conversations.

Operation Glowing Symphony

Around 2015, US military intelligence was given the directive to counter the online presence of the terrorist group Islamic State of Iraq and Syria (ISIS), which was using websites and social media to attract recruits around the world. US intelligence had to adapt to an enemy that was skilled at stoking hatred against America and its Western allies on the internet.

The Americans began hacking ISIS's online networks, collecting IP addresses and passwords. Then in 2016 and 2017, the military conducted a successful cyberwarfare operation called Operation Glowing Symphony to disrupt the online media and propaganda channels of the terrorist organization. AI aided analysts in processing huge amounts of data that illustrated how ISIS's networks operated. The hackers gathered ISIS videos, financial records, and other information from servers all over the world. Then the United States locked key ISIS members out of the servers used to spread propaganda.

Operation Glowing Symphony was a success. ISIS internet communications were disrupted, and large amounts of their recruitment information on the internet disappeared. In Operation Glowing Symphony, the AI technologies automated tasks that would have been impossible for humans to achieve using traditional methods.

messages to each other, with each reading their message in their own language.

In 2001 the US military began using a handheld device called the Phraselator. This system helped troops who spoke different languages handle basic communication when a human interpreter was not present. Over time, this technology has evolved. By 2017, US troops were using the more sophisticated Machine Foreign Language Translation System, which employs machine learning to provide US troops with speech-to-speech and text-to-text translations. A US Army news release described the system, saying, "The real magic of the Machine Foreign Language Translation System lives deep within the language packs, where science and art come together to enable software to hear, understand and interpret as much like a human linguist as possible."[29] Today AI allows the US military to communicate with its allies in real time. It also allows the US military to analyze large amounts of intelligence originating in countries that are not US allies and detect potential threats.

In early 2024 more than ninety thousand troops from all thirty-two NATO countries took part in the Steadfast Defender 24 exercise. Language translations were a necessary part of this operation, and AI-powered machine translation helped overcome the language barriers. It paved the way for smooth communication between troops from the various participating countries.

AI in Cybersecurity

The US military is also using AI in cybersecurity, protecting its computer networks and devices from digital attacks by enemies and hackers. As computer technology has grown more sophisticated, the cyberattacks against these systems have become cleverer and more difficult to detect. The military is using AI to monitor internet traffic and flag potential threats, identify behavior patterns that might show the enemy's plans, and help military personnel in security decision-making.

The military uses a zero-trust framework for its computer networks. This is a computer security model focused on the principle "never trust, always verify." Older traditional security models assumed that the users and devices within the same network were trustworthy. The newer zero-trust framework assumes that all users, devices, and applications cannot be automatically trusted and must be verified first. The US military is using AI-powered automation and a zero-trust framework to protect its networks against cyberattacks.

Under the zero-trust framework, AI technology continuously monitors the computer networks, verifying that users, devices, and applications are trustworthy. Software bots search for unusual activity on any network and identify potential threats. The automatic software bots powered by AI technology also handle software updates on the military networks. Up-to-date software helps keep the networks more secure. With software bots handling the necessary updates automatically, military personnel can focus their attention on more complex security tasks.

The US military uses a zero-trust framework to protect its networks against cyberattacks. Under the zero-trust framework, AI technology continuously monitors the computer networks and users.

AI in Predictive Analytics

Just as good chess players are skilled at predicting their opponents' next moves, military intelligence agencies strive to foresee their enemies' next moves on the battlefield or the world stage. With predictive analytics, the military uses data to forecast future outcomes. Algorithms and machine learning help identify patterns in the data and suggest logical outcomes.

Traditionally, militaries have done their work in a reactive way. When a crisis breaks out somewhere in the world, the military reacts by sending troops and supplies to critical locations. But with predictive analytics, militaries can be much more proactive, positioning troops and supplies in the right place to perhaps prevent the crisis from occurring or escalating.

The US Army uses the Persistent Surveillance and Dissemination System of Systems (PSDS2) to locate enemies and

"Predictive analytics . . . offers a way to control the tempo of operations by forcing the enemy into a series of reactive moves, all while keeping one step ahead."[30]

—Anthony Grajales, US Army major

predict their movements. The PSDS2 collects data from a wide range of sources, including drones, balloons, and ground-based radar. The system uses machine learning algorithms to sort the data and provide an "operational picture" that can easily be monitored. It can then offer predictive models of an enemy's activities.

Monitoring enemy activities and forecasting outcomes provides soldiers and their commanders with quicker decision-making capabilities. Major Anthony Grajales described the potential of predictive analysis in a 2025 US Army news release:

> Predictive analytics, in this sense, becomes a tool for out-maneuvering rather than simply outpacing. It offers a way to control the tempo of operations by forcing the enemy into a series of reactive moves, all while keeping one step ahead. . . . As military logistics continues to evolve, the use of predictive analytics will only grow more sophisticated. We can envision a future where logistics networks are so finely tuned that they can respond to shifting conditions in real-time, preemptively rerouting supplies, repositioning personnel, and dynamically adjusting strategies based on the latest data inputs. This is made possible by advancements in algorithms that change and create running estimates more quickly to enhance decision-making processes. Imagine running the military decision-making process in minutes thanks to predictive analytics and AI. This is a serious upgrade for decision making.[30]

With AI technologies, the military can gather information about its enemies and process this information more quickly than ever before. Knowing where the enemy is located and what the enemy is doing gives troops a decided advantage in warfare.

AI in Military Efficiency

The US military has the distinctive mission to support and defend the Constitution against all its enemies. To carry out that duty, the military must be well organized, maintaining command over its troops while ensuring that all soldiers, sailors, and air personnel are compensated for the vital work they do. To some, the military seems to operate like a very large business. With a workforce of about 2.9 million, there is much that goes on behind the scenes to keep the "business" of the military running smoothly. Troops are trained. Paychecks are issued. Supplies are moved.

AI Supplies the Troops

With troops working around the world at bases on land and aboard ships at sea, acquiring, moving, and distributing supplies is a complex job. To streamline these tasks, the military uses AI tools. AI supply-chain programs help staff members analyze data so they purchase the right amount of supplies and plan the most efficient schedules and travel routes for moving those supplies to the troops. The added efficiency saves time and money.

The Defense Logistics Agency manages the US military's supply chain, providing troops with important items like uniforms, fuel, food, tools, and repair parts. The Defense Logistics Agency began using AI in 2018 to increase its efficiency. The first AI tool adopted was a virtual assistant called Val. Agency staff used Val to help solve information technology problems. Today the agency is using newer AI tools to do more things.

Ruksana Lodi is the AI officer for the Defense Logistics Agency. She explains how AI helps the logistics team avoid purchasing duplicate supplies. "If you go out and find a tool on your own and buy it, and then I do the same, we would end up with many tools that may have the same functionality," Lodi says. AI helps employees plan for and purchase supplies more efficiently. "For example, AI can be used to prevent stockouts and overstocking. This keeps supplies available when needed, avoids delays and reduces inventory costs,"[31] says Lodi. AI is helping the military lower costs and move supplies more efficiently.

AI Predicts Future Needs

AI helps the military operate more efficiently in other ways, too. For instance, AI can predict when equipment will break down or when parts will need routine service. The US military has numerous tanks, ships, aircraft, trucks, helicopters, and weapons, and all must be kept in good working order. The military uses predic-

A military vehicle is loaded for transport by a Defense Logistics Agency team. The agency began using AI in 2018 to make it more efficient in managing the military supply chain.

tive maintenance powered by AI to ensure its equipment runs well. In many cases, the military is able to maintain and repair this equipment before it breaks down, saving time and money.

> "Predictive analytics can monitor equipment health, usage patterns, and environmental factors to anticipate failures and schedule maintenance proactively. This minimizes downtime, extends equipment lifespan, and ensures maximum operational availability."[32]
>
> —Christopher O. Mohan, lieutenant general

The path to successful predictive maintenance begins when sensors are placed in equipment and vehicles. These sensors relay real-time data that is analyzed by AI algorithms. The algorithms sort through patterns in the data, looking for anomalies that suggest equipment might fail. The AI can cross-reference this information with maintenance records and then recommend a plan for repair. Lieutenant General Christopher O. Mohan says:

> Predictive analytics can monitor equipment health, usage patterns, and environmental factors to anticipate failures and schedule maintenance proactively. This minimizes downtime, extends equipment lifespan, and ensures maximum operational availability. The Army's fleet of vehicles and aircraft can benefit significantly from predictive maintenance, enhancing readiness and reducing sustainment costs.[32]

For the US Navy, AI-powered predictive maintenance can help pinpoint a faulty system on a ship or a submarine before it becomes a serious problem. Maintenance crews with the skill and tools to address the issue can fix the system at sea or in the next port if the fault is more significant. If the problem needs immediate attention but the required part or tools are unavailable, the Navy might fly a maintenance crew to a ship at sea with the correct repair parts to make the fix.

Preventing Accidents

Like the Navy, the Army began using predictive maintenance for its AH-64 Apache helicopters in 2005. A few years later the Army applied the process to its fleet of UH-60 Black Hawk helicopters, too. By 2022 it was also using predictive maintenance with its CH-47 Chinook helicopters. Before using predictive maintenance, it was a challenge for helicopter mechanics to find and fix problems before they occurred. If they tried to swap out worn parts early, they risked removing parts that were still useful. With predictive maintenance, the Army is fixing equipment before it breaks and becomes a much bigger problem.

In one case, the Army used predictive maintenance AI with a UH-60 Black Hawk helicopter. The helicopter had unusual tail rotor vibrations. With predictive maintenance, the Army knew that the helicopter's gearbox was failing. "During the teardown inspection, we determined that this one was within hours of catastrophic failure, which would have likely resulted in loss of aircraft," said Major General Thomas O'Connor in a 2023 interview. Predictive maintenance is helping the military prevent accidents from happening, while also saving money. "Every day we are getting these red flags. We're removing components and replacing them earlier in order to save short-term readiness and prevent any catastrophic failures,"[33] said O'Connor.

The Army uses predictive maintenance AI to help it manage its UH-60 Black Hawk helicopters (pictured). With predictive maintenance they can fix equipment before it breaks.

AI Finds Space Junk

The very newest military branch, the US Space Force, originated in 2019. Its mission, as stated on the Space Force website, is to "secure our nation's interests in, from, and to space." The Space Force is using AI technology to fulfill its mission. This includes AI tools that process the data, allowing military personnel to track and identify objects that are in orbit, including satellites and debris. If the Space Force knows where the debris is, it can move its satellites to safer orbits. The debris, or space junk, includes satellites that are no longer functioning and pieces of rockets that have served their purpose but are still orbiting Earth. It can even include small debris, like chips of paint, tools, or other things spacecraft might leave behind. The National Aeronautics and Space Administration reports that there are more than twenty-five thousand debris objects larger than 3.9 inches (10 cm) orbiting Earth.

Space debris is a problem because these orbiting pieces of junk travel at extremely high speeds and can damage active satellites. These working satellites are providing navigation, communications, and weather data for humans on Earth. "The Space Force recognizes the critical role that data and artificial intelligence will play in maintaining space superiority," says Colonel Nathen Iven.

US Space Force, "United States Space Force: About Us." www.spaceforce.mil.

Quoted in Sandra Erwin, "Space Force Unveils Strategic Plan for AI Integration," Space News, March 19, 2025. https://spacenews.com.

Predictive Logistics

The US military is also using AI-powered predictive analysis in logistics. AI helps military personnel place the troops and supplies precisely where they are needed at the right time. Before predictive logistics was implemented, the military would see where troops or supplies were needed and then react by sending the personnel and supplies to that location. Sometimes, this would mean delays because soldiers and cargo had to prepare for transport, and either one could be in short supply. "Predictive logistics represents a shift from traditional, reactive sustainment models to a proactive, data-driven approach that allows us to position supplies to ensure the right resources are available at the right time and place,"[34] says Mohan.

Predictive logistics works with AI algorithms that analyze historical data along with the military's current plans and needs. "For example, the Army can anticipate the demand for specific munitions based on emerging threats and operational requirements, optimizing inventory levels,"[35] says Mohan. Because the military knows when and where supplies and personnel will be needed, it can ensure that shortages do not occur.

AI in Training

Before the military sends troops into battle, though, those troops have spent countless hours training for combat and survival. Today, the US military is integrating AI into its training programs in a variety of ways.

One program is helping soldiers become more accurate in their shooting and decision-making. In the past soldiers used training videos. The videos were displayed on a large screen, and the soldier used a special weapon to fire lasers at targets. But these training scenarios were unchanging, so there was little to be learned or gained after a soldier had practiced them repeatedly.

The AI-powered VirTra simulation training program at Fort Carson in Colorado now provides soldiers with video game–like scenarios that are focused on each individual soldier's performance and needs. VirTra uses virtual reality gear to create immersive environments that a soldier can roam through when attempting to complete a training mission. The AI evaluates a soldier's progress in the training and modifies the scenarios to create new challenges. With AI, the training can be specifically tailored to each soldier's needs.

VirTra also includes physical exercises during the training to elevate the soldier's heart rate, mirroring the excitement and anxiety encountered in actual combat. The VirTra training scenarios might involve sending soldiers through urban environments that include virtual civilians, and the users will have to make quick decisions about when and where to shoot their virtual gun if force is needed. "This allows us to pair marksmanship with a variety of different training environments where they are coupled with a strength coach to simulate physical stress,"[36] says Jake Blumberg, a cognitive performance specialist who supervised the

AI Helps Match Jobs

In 2024 the US military launched GigEagle, a project that uses AI to match personnel having specific skills with important short-term—or gig—jobs. This tool is especially valuable for National Guard and Reserve troops who serve on a part-time basis. When not on duty, they hold civilian jobs. The National Guard and Reserve troops can be called for full-time active duty in a national emergency.

With GigEagle, the National Guard and Reserve troops build a profile highlighting their expertise. When called to duty, the program identifies a reservist's skills and connects that individual to a critical need. For example, when Hurricane Milton hit Florida in 2024, the military used GigEagle to find National Guard and Reserve members with skills such as civil engineering or truck driving to assist in storm recovery.

The system also helps align reservists with expertise that might be needed in combat. "What we're doing is unlocking a supply of talent that has previously been inaccessible. When I talk about reestablishing deterrence, that's what I mean—because now I have unlocked the ability to engage that expert in radio frequencies at Caltech or MIT, or somebody at UT [the University of Texas at] Austin who is proficient in designing drones," says Air Force Reserve Brigadier General Michael McGinley.

Quoted in Anastasia Obis, "GigEagle Lays Groundwork for Whole-of-Nation Talent Marketplace," Federal News Network, June 13, 2025. https://federalnewsnetwork.com.

training. The Tenth Special Forces Airborne Group, also known as the Green Berets, completed this training in 2024.

The military has other AI-powered training tools. The Personal Assistant for Life Long Learning (PAL3) is an AI tutor that sharpens educational skills. The US Navy used early versions of PAL3 help sailors in electronics-technicians roles keep their expertise up-to-date. Today the military is using it to help soldiers stay proficient in using AI.

AI in War Games

Computer or tabletop war games also train military personnel to make strategic decisions on the battlefield. The games model actual battle scenarios, and they offer multiple paths to victory or defeat. The US military has used war games for more than one hundred years to role-play the decision-making that happens in real battles. The simulations give the troops realistic missions and challenges so that they gain experience.

In recent years, AI has been added to the data analysis portion of these simulations. Generative AI analyzes the success or failure of previous players, modifying new scenarios to keep the experience fresh for recruits. With the addition of AI, the war game scenarios can be greatly expanded, providing military personnel with millions of battle scenarios in a short period of time.

The Schriever Wargame is one example of the US military's stockpile of scenarios. The Schriever Wargame began in 1998 and has been held regularly since then. It is a tabletop game now hosted by the US Space Force to train for conflicts in futuristic environments, including outer space. In the 2025 playthrough, the United States collaborated with nine other nations to wargame a conflict set in the Indo-Pacific region ten years in the future. The Indo-Pacific region covers the lands and waters around the Pacific and Indian Oceans. The game centered on probable technologies and potential conflicts that might arise in 2035. The 2025 Schriever Wargame focused on the question of how the United States and its allies will conduct operations in space, on land, and at sea as the region becomes more congested.

> "Generative AI will fundamentally reshape war gaming . . . by allowing senior military and political leaders to pursue better tactical solutions to unexpected crises, solve more complex logistical and operational challenges and deepen their strategic thinking."[37]
>
> —John Long Burnham, senior policy research analyst at the University of Alberta

All branches of the US military use war games in their training. "Generative AI will fundamentally reshape war gaming . . . by allowing senior military and political leaders to pursue better tactical solutions to unexpected crises, solve more complex logistical and operational challenges and deepen their strategic thinking,"[37] says John Long Burnham, a policy analyst, in a Military.com article on war gaming.

The Johns Hopkins University Applied Physics Laboratory (APL) in Maryland is exploring how AI can automate the creation of military war games. APL is applying large language models to military war

In the 2025 Schriever Wargame (pictured), the United States collaborated with nine other nations to war-game a mock conflict set in the Indo-Pacific region ten years in the future.

games. Large language models are AI systems that are trained to understand and generate human language. ChatGPT is an example of a chatbot that draws from a large language model. APL's large language model capability could do the job of the human analysts who design, set up, develop, and run the war games. "This work is a major step forward in wargaming," says Andrew Mara, head of APL's National Security Analysis Department. "Senior leaders in the Department of Defense have been seeking something like this for more than a decade, and I think we are finally at the point where the demand signal and the technology are in alignment."[38]

The APL technology could help make war games more accessible to military personnel everywhere. Kevin Mather, an analyst who is leading this work at APL, says:

> Our long-term vision is to create a tool that allows decision-makers at many different levels to arrive at the kind of insights that come from wargaming in a form that lives on their personal computers and allows them to run

> a lot of repetitions in a short period of time, instead of a full-blown exercise that takes months of planning. Of course, there are advantages that the players get from in-person exercises that you don't get from a purely digital format, but the idea is to provide more options.[39]

From the battlefield to the classroom, AI is changing the ways the military does its job. AI helps the military work more efficiently. With its predictive powers, AI can alert maintenance crews before a piece of equipment breaks. It can enhance the mental and physical training soldiers must undergo. And it can keep timetables that ensure supplies and troops reach critical areas on the modern battlefield. Altogether, these capabilities show how AI is transforming not just the tools of warfare but also the very decision-making processes that define military operations.

SOURCE NOTES

Introduction: AI Finds the Target

1. Quoted in Joe Saballa, "US Military Deploys AI to Aid Air Strikes in Middle East," Defense Post, March 6, 2024. www.thedefensepost.com.
2. Quoted in Secretary of the Air Force Public Affairs, "DAF Leaders Emphasize Adapting AI for Warfighting Success," December 2, 2023. www.af.mil.
3. Kathleen Hicks, "AI in the Era of Strategic Competition," US Department of Defense, March 5, 2024. www.defense.gov.

Chapter One: Seeking the AI Advantage

4. Kathleen Hicks, "The State of AI in the Department of Defense," US Department of Defense, November 2, 2023. www.defense.gov.
5. Mark T. Esper, "Remarks by Secretary Esper at National Security Commission on Artificial Intelligence Public Conference," US Department of Defense, November 5, 2019. www.defense.gov.
6. Sam J. Tangredi, "Sun Tzu Versus AI: Why Artificial Intelligence Can Fail in Great Power Conflict," US Naval Institute, May 2021. www.usni.org.
7. DARPA, "About DARPA: Our Mission." www.darpa.mil.
8. Quoted in Julian Turner, "Sea Hunter: Inside the US Navy's Autonomous Submarine Tracking Vessel," Naval Technology, May 3, 2018. www.naval-technology.com.
9. Jake Sullivan, "Remarks by APNSA Jake Sullivan on AI and National Security," White House, October 24, 2024. www.bidenwhitehouse.archives.gov.
10. Quoted in Brit McCandless Farmer, "AI in the Military: Gen. Milley on the Future of Warfare," CBS News, October 8, 2023. www.cbsnews.com.

Chapter Two: AI in Battle

11. Hicks, "The State of AI in the Department of Defense."
12. Quoted in Katrina Manson, "AI Warfare Is Already Here," Bloomberg Businessweek, February 28, 2024. www.bloomberg.com.

13. Clifford A. Baxter, “Enhancing Tactical Level Targeting with Artificial Intelligence,” Line of Departure, March 1, 2024. www.lineofdeparture.army.mil.
14. Baxter, “Enhancing Tactical Level Targeting with Artificial Intelligence.”
15. Quoted in Noah Haggerty, “Using AI to Save Lives on the Battlefield,” Northeastern University College of Engineering, July 8, 2024. https://coe.northeastern.edu.
16. Quoted in Andrew Salerno-Garthwaite, “Project ATRACT—War-Zone Triage with AI and UAV,” Army Technology, March 21, 2023. www.army-technology.com.

Chapter Three: AI in Aircraft, Weapons, and Ships

17. Quoted in Laura Gozzi, “How Ukraine Carried Out Daring ‘Spider Web’ Attack on Russian Bombers,” BBC, June 2, 2025. www.bbc.com.
18. Quoted in Haley Britzky and Isabelle Khurshudyan, “US Drone Dilemma: Why the Most Advanced Military in the World Is Playing Catchup on the Modern Battlefield,” CNN, September 15, 2025. www.cnn.com.
19. Quoted in Britzky and Khurshudyan, “US Drone Dilemma.”
20. Quoted in Samuel King Jr., “Autonomous F-16 Program Advances with Modifications, Simulations,” US Air Force, April 2, 2025. www.af.mil.
21. Quoted in King, “Autonomous F-16 Program Advances with Modifications, Simulations.”
22. Quoted in UN News, “‘Politically Unacceptable, Morally Repugnant’: UN Chief Calls for Global Ban on ‘Killer Robots,’” May 14, 2025. https://news.un.org.
23. Quoted in DARPA, “DARPA Christens Unmanned Ship Aimed at Revolutionizing Naval Capability,” August 11, 2025. www.darpa.mil.
24. Quoted in Aaron-Matthew Lariosa, “Serco, DARPA Launch NOMARS USX-1 *Defiant*,” Naval News, April 3, 2025. www.navalnews.com.

Chapter Four: AI in Intelligence Gathering

25. Quoted in Gareth Jennings, “Sweden to Field Fast-Tracked ‘Swarming Drone’ Capability, Says Defence Minister,” Janes, January 15, 2025. www.janes.com.

26. Hicks, "The State of AI in the Department of Defense."
27. Quoted in Greg Kocher, "A New Tool for Intruder Detection: BGAD Tests AI for Enhanced Security," US Army, October 8, 2024. www.army.mil.
28. Quoted in Wes Shinego, "Defense Department Tests AI Software, Advances to Improve Physical Security Posture," Department of Defense Manufacturing Technology Program, October 28, 2024. www.dodmantech.mil.
29. Quoted in Michael Doney et al., "Technically Speaking: Making Language Less Foreign," US Army, October 17, 2017. www.army.mil.
30. Anthony Grajales, "The Predictive Turn: Preparing to Outthink Adversaries Through Predictive Analytics," US Army, January 22, 2025. www.army.mil.

Chapter Five: AI in Military Efficiency

31. Quoted in Beth Reece, "AI to Boost Efficiency, Optimize Logistics Support as DLA Standardizes Use of New Tech," Defense Logistics Agency, March 17, 2025. www.dla.mil.
32. Christopher O. Mohan, "Predictive Logistics Is the Way of the Future," US Army, January 22, 2025. www.army.mil.
33. Quoted in Jen Judson, "US Army Turns to Predictive Maintenance to Cut Mishaps," Defense News, January 19, 2023. www.defensenews.com.
34. Mohan, "Predictive Logistics Is the Way of the Future."
35. Mohan, "Predictive Logistics Is the Way of the Future."
36. Quoted in John Breeden II, "Examining Some Places Where AI Has Already Been Deployed Within the Military," Nextgov, February 26, 2024. www.nextgov.com.
37. John Long Burnham, "The Use of AI in War Games Could Change Military Strategy," Military.com, April 22, 2024. www.military.com.
38. Quoted in Ajai Raj, "Generative AI Wargaming Promises to Accelerate Mission Analysis," Johns Hopkins Applied Physics Laboratory, March 3, 2025. www.jhuapl.edu.
39. Quoted in Raj, "Generative AI Wargaming Promises to Accelerate Mission Analysis."

FOR FURTHER RESEARCH

Books

John Allen, *Artificial Intelligence: Promise and Peril*. ReferencePoint, 2024.

Denise Garcia, *The AI Military Race: Common Good Governance in the Age of Artificial Intelligence*. Oxford University Press, 2024.

Avery Elizabeth Hurt, *Opposing Viewpoints: Artificial Intelligence*. Greenhaven, 2023.

Carla Mooney, *Thinking Critically: Artificial Intelligence*. ReferencePoint, 2025.

Stephanie Sammartino McPherson, *Future Forward: Harnessing Artificial Intelligence*. Twenty-First Century, 2025.

Scientific American, ed., *Scientific American Explores Big Ideas: Artificial Intelligence.* Rosen, 2023.

Internet Sources

BBC, "What Is AI, How Does It Work and Why Are Some People Concerned About It?," July 29, 2025. www.bbc.com.

Joshua Glonek, "The Coming Military AI Revolution," *Military Review*, May–June 2024. www.armyupress.army.mil.

David Kirichenko, "Artificial Intelligence's Growing Role in Modern Warfare," *War Room*, August 21, 2025. https://warroom.armywarcollege.edu.

Sentient Digital, "The Most Useful Military Applications of AI in 2024 and Beyond," March 15, 2024. https://sdi.ai.

Cole Stryker and Eda Kavlakoglu, "What Is Artificial Intelligence (AI)?," IBM. www.ibm.com.

Websites

Chief Digital and Artificial Intelligence Office
www.ai.mil
This Department of Defense website has information about the military's adoption of AI in its operations.

Future of Life Institute
https://futureoflife.org
The Future of Life Institute is a nonprofit organization focused on seeing transformative technology used in ways that benefit, rather than harm, humanity. The website has articles on emerging technologies, such as artificial intelligence and the military.

Military AI
https://militaryai.ai
This website is part of Defense Post, an independent media outlet that focuses on global defense-related news. Military AI provides news stories related to the military's use of artificial intelligence.

INDEX

Note: Boldface page numbers indicate illustrations.

PICTURE CREDITS

Cover: U.S. Navy photo by Ensign Pierson Hawkins

5: Vblinov/Shutterstock
9: Gorodenkoff/Shutterstock
13: DenPhotos/Shutterstock
15: U.S. Navy photo by Ensign Pierson Hawkins
18: Z2A Collection/Alamy Images
20: Andrei Armiagov/Shutterstock
24: World History Archive/Newscom
28: hurricanhank/Shutterstock
31: Faizinraz/Shutterstock
33: A.PAES/Shutterstock
36: Mike Mareen/Shutterstock
40: Gorodenkoff/Shutterstock
43: M Agency/Shutterstock
46: NB/ROD/Alamy Images
48: WildWestScott/Shutterstock
53: Billy Blankenship

ABOUT THE AUTHOR

Donna B. McKinney is the author of more than thirty-five nonfiction books for young people on topics that include science, technology, history, sports, and current events. Before McKinney wrote for young people, she spent many years writing about military science and technology topics, such as chemistry, space science, and robotics, at the US Naval Research Laboratory in Washington, DC. She has a BA in English from Campbell University and an MA in English from George Mason University. She lives in North Carolina.